RADAR and HUMAN VISION:
Comparison Hypothesis for
UFOLOGY

H. David Umphenour

Copyright © 2018 by H. David Umphenour

All rights reserved, including the right of reproduction in whole or in part in any form.

ISBN: 1727469658
ISBN-13: 978-1727469653

DEDICATION

To knowledge at any level, wisdom at any age and Betsy.

CONTENTS

	Epigraphs	i
1	Introduction	1
2	The UFO	3
3	The Surface Properties	4
4	The Geometry	5
5	The UFO Ambient Reflected Radiation	6
6	The Troposphere	7
7	The Radiometry	8
8	Human Vision	10
9	RADAR	13
10	The UFO Thermal Emitted Radiation	17
11	Summary	19
12	Works Cited	21
13	Index	23
	The Author	25

EPIGRAPHS

Your theory is crazy but it's not crazy enough to be true. - N. Bohr

Theories have four stages of acceptance: 1) this is nonsense; 2) this is interesting, but a weird point of view; 3) this is true but insignificant; 4) I always said so. - J.B.S. Haldane (paraphrased)

An extraordinary phenomenon demands an extraordinary investigation. - L. Kean

---

Where is everybody? - E. Fermi

Doubt everything or believe everything: these are two equally convenient strategies. With either we remove the need for reflection. - H. Poincaré (paraphrased)

My own conclusion is that they don't exist. - R. Kurzweil

1 INTRODUCTION

This is not about conspiracy or encounters or weird craft or life-forms. It's about applying what is known to the unknown. Most of ufology is bogus, but within this context, ufology is, from the Greek Logos, non-sensationalized ordering principles (words) about UFOs (Unidentified Flying Objects, Unidentified Aerial Phenomena, etc.) whatever they may or may not be. The author neither confirms, nor denies, their existence. The text is also an introduction to RADAR (RAdio Detection And Ranging) and Human Vision. It addresses a quantifiable aspect of ufology that appears to have a recurring presence that has yet to reveal its true nature. The fact that it's recurring is what makes it interesting!

Since the middle of the twentieth century, the so-called UFO modern era, one can find consistent survey literature and study statements like the following sentence. "Perhaps five to ten (or higher) per cent of investigated sightings cannot be explained by common-sense, known phenomenology or present scientific and technological understanding." If reality reflects the truth, that's amazing! The review of the specious Condon Report by the American Institute of Aeronautics and Astronautics in the late 1960s (Kean) is an early example of many.

Philosophers of science opine something to the effect "Ufology is a pseudoscience" while in the next breath stating something trivial like, "SETI (Search for Extra-Terrestrial Intelligence) is a young developing science". More power to SETI, but no, the hard data content so far for both ET intelligence and life is zero. Plus, SETI's theoretical foundations are marginal at best (Pigliucci). The Drake equation with its wild fractional probabilities (or Kardashev's et al.) is hardly a foundational theory. It's no accident that it's called Search for ETI, rather than Science of ETI.

Columbus was a searcher not a scientist. What he stumbled across a few hundred miles off the Florida coast was not what he expected. What if

SETI stumbles across something it's not expecting? How would they interpret it? When J. Bell and A. Hewish discovered what later became known as the first pulsar, the idea that it could be of ET origin was briefly entertained. If a credible UFO sighting were to yield an iota of anything new or paradigmatic, isn't that infinitely more than SETI has produced, or is likely to produce?

Yes, the author has digressed, back to the topic of interest. Another generic statement found in the community goes something like the following sentence. "Visual and RADAR corroboration is desirable when investigating a credible sighting." OK. First, some questions. Is this a practical expectation? Under what conditions is it possible or impossible? Probably the most important, does either detection mode or their corroboration inform one about the nature of the UFO? So... the author looks to the 1949 physics Nobelist H. Yukawa regarding that nature (Wilson). "Suppose there is something which a person cannot understand. He happens to notice the similarity of this something to some other thing which he understands quite well. By comparing them he may come to understand the thing which he could not understand up to that moment..."

In the next chapter an example is launched using Yukawa's approach with theories, laws and results from physical and geophysical science, RADAR engineering and human psychophysics, i.e., how the eye/brain responds to a stimulus like light (Blackwell, Kahneman).

2 THE UFO

The object will be a metallic sphere, such as Aluminum or Copper, with radius r, exhibiting a relevant RADAR cross-section. No exotic RADAR altering material or structure as used on the F-117 fighter or B-2 bomber are considered. The sphere may be covered with non-black military paints or coatings (Fox). RADAR details are considered later. Human vision is considered first. From this point on in the development the sphere is referred to as the UFO. The reader should not infer from this definition that there can be no other physical or phenomenology aspect.

3 THE SURFACE PROPERTIES

Using Kirchoff's law, the diffuse hemispherical emittance ε_H, diffuse hemispherical reflectance ρ_H and specular reflectance ρ_s (Dudzik) of the opaque UFO are related as

$$\epsilon_H = 1 - \rho_H - \rho_s$$

and will be fixed at ~1/5=20% (Wolfe), ~1/8=13% (Fox) and ~2/3=67% (Wolfe) respectively for the UFO at the peak of all vision spectra. The reflected specular component only occurs within a lobe or cone with apex half-angle φ, i.e., solid angle $\pi(\sin(\varphi))^2$ (Dudzik). If a specular reflection occurs, it will happen at a surface normal to the UFO that is 7.5 degrees below the horizontal. The Sun will be incident at 22.5 degrees above normal with the center of the specular cone reflected at 22.5 degrees below normal. The photosphere of the Sun on the incident side subtends a circular diameter of 0.53 degrees from Earth (Jursa). The specular reflection cone is assumed to be equal to that from the incident Sun or

$$\pi(\sin(0.265\deg))^2 \cong 67 \mu sr$$

which is less than that of most terrestrial reflecting materials (Dudzik). The distance d from the specular reflection to the observation point is

$$d = -0.924r + (R^2 - 0.146r^2)^{1/2} \cong R$$

where R is the distance from the center of the UFO to the observation point.

4 THE GEOMETRY

The UFO is at 30 degrees elevation above the observation point on Earth. The Sun is 180 degrees in azimuth from the UFO at an elevation of 15 degrees above the observation point. The human observer looks at the UFO and is back-lit by the Sun. The vertical plane containing the UFO, observer and Sun defines where all vision and RADAR calculations are made. Similarly, night-time conditions are in the same plane without any Moon phase above the horizon. Day and night are clear with no clouds. The view diagram is the vertical plane from the perpendicular direction.

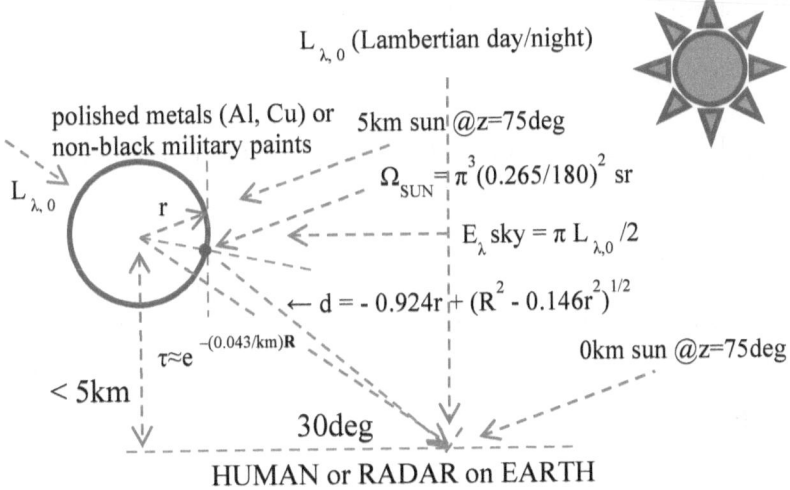

5 THE UFO AMBIENT REFLECTED RADIATION

The UFO will reflect light in the visible from two sources. Those are: direct solar radiation from the Sun and atmospheric scattered radiation. The latter is sometimes called sky-shine and can be a result of solar scattering in the day or scattering of starlight at night. There are tertiary and higher levels involving ground reflections of these components, but for this scenario they are deemed insignificant. Polarization effects are not considered.

The spectral irradiance at the day-time vision spectral peak for direct solar radiation at a 15 degree elevation can be expressed as a function of altitude from h=0 to 5km in the troposphere as (Zissis)

$$E_{\lambda,sun} = 60 mW/cm^2/\mu m + (11 mW/cm^2/\mu m/km)h.$$

The Lambertian spectral radiance of the sky in the lower troposphere at the day-time vision spectral peak is (Zissis)

$$L_{\lambda,0} = 1.5 mW/cm^2/\mu m/sr.$$

The same quantity at the night-time vision spectral peak due to starlight is (RCA)

$$L_{\lambda,0} = 0.4 nW/cm^2/\mu m/sr$$

or three and three-quarter million times lower than day-time. Lambertian means the radiance of an object is constant with respect to the aspect from which it is viewed.

6 THE TROPOSPHERE

The attenuation by the atmosphere for both the day-time and night-time vision spectral peaks will result from aerosol and molecular Rayleigh scattering with minimal ozone, water vapor and carbon dioxide absorption, producing a transmission given by (RCA, Zissis)

$$\tau \cong e^{-(0.043/km)R}.$$

This is approximately equivalent to a 20km horizontal visibility at sea-level (RCA). Optical effects due to turbulence are not considered.

7 THE RADIOMETRY

Normally, visible light implies photometric notation. However, the author's preference is to proceed with an equivalent radiometric development (e.g., Dudzik) and convert to photometric units (e.g., RCA) as needed.

In the day-time the Sun will illuminate a portion of the UFO whose projected area viewed from the observation point is part ellipse and part semicircle or

$$\left(\frac{2^{1/2}}{4} + \frac{1}{2}\right)\pi r^2.$$

The spectral solar irradiance normal to this projected area is

$E_{\lambda,sun} \cos(45deg).$

The spectral sky irradiance from the quarter-sphere of atmosphere on the observer side of the UFO is approximately

$$\frac{\pi}{2} L_{\lambda,0} = \pi L_{\lambda,0} \int_0^{90deg} \cos\theta \sin\theta \, d\theta$$

with that normal to the UFO projected area as

$$\frac{\pi}{2} L_{\lambda,0} \cos(30deg).$$

The diffuse reflected spectral radiance from the UFO as viewed from the observation point corrected for atmospheric transmission involves the sum of these components such that

$$L_\lambda = \left(\left(\frac{2^{1/2}}{4} + \frac{1}{4}\right)E_{\lambda,sun} + \frac{3^{1/2}\pi}{4}L_{\lambda,0}\right)\frac{\rho_H}{\pi}e^{-(0.043/km)R}.$$

To this the corrected specular reflected spectral radiance from the UFO is added giving

$$L_\lambda = \left(\left(\frac{2^{1/2}}{4} + \frac{1}{4}\right)E_{\lambda,sun} + \frac{3^{1/2}\pi}{4}L_{\lambda,0}\right)\frac{\rho_H}{\pi}e^{-(0.043/km)R} + \left(\frac{E_{\lambda,sun}}{\pi(\sin\phi)^2} + L_{\lambda,0}\right)\rho_s e^{-(0.043/km)d}.$$

This is the total reflected spectral radiance of the UFO as seen from the observation point. If the direct Sun is removed as a source then

$$L_\lambda = \frac{3^{1/2}}{4}L_{\lambda,0}\rho_H e^{-(0.043/km)R} + L_{\lambda,0}\rho_s e^{-(0.043/km)d}.$$

The contrast as viewed from the observation point is defined as (Blackwell)

$$C = \frac{L_\lambda}{L_{\lambda,0}} - 1.$$

The day-time contrast then is

$$C_D = \left(\left(\frac{2^{1/2}}{4} + \frac{1}{4}\right)\frac{E_{\lambda,sun}}{L_{\lambda,0}} + \frac{3^{1/2}\pi}{4}\right)\frac{\rho_H}{\pi}e^{-(0.043/km)R} + \left(\frac{E_{\lambda,sun}}{L_{\lambda,0}\pi(\sin\phi)^2} + 1\right)\rho_s e^{-(0.043/km)d} - 1$$

with the night-time contrast

$$C_N = \frac{3^{1/2}}{4}\rho_H e^{-(0.043/km)R} + \rho_s e^{-(0.043/km)d} - 1.$$

All specular components in the reflected spectral radiance and contrast can be removed by setting the specular reflectance to zero.

8 HUMAN VISION

The threshold contrast levels for 50% probability of detection with unlimited observation time for unaided ~20/20 human vision are determined from the classic study of Blackwell. Threshold contrast in general is a function of background brightness and the angular sub-tense θ of the object viewed. For moderate changes in day-time sky brightness, the threshold contrast approaches a small constant value as θ becomes large. This behavior is referred to as the Weber-Fechner law (Blackwell, Kahneman).

In the day-time cases, photopic cone vision will occur with a peak response at 0.555 micron wavelength (green approaching yellow). At night, scotopic rod vision (i.e., color vision loss) occurs with a peak response at 0.51 micron wavelength (green approaching blue). For this study the Full-Width-Half-Maximum of vision response is assumed to be at plus and minus 0.05 micron (FWHM=0.10 micron) for both peaks (RCA).

The day and night radiometric background radiance (watts per square meter per steradian) can be converted to photometric background brightness (candela per square meter) using (RCA)

$$600,000 cd/m^2 = 892 W/m^2/sr \; photopic$$

or

$$600,000 cd/m^2 = 348 W/m^2/sr \; scotopic.$$

It should be noted that it is (was) standard in photometry to work in the photopic conversion space. This work, as well as Blackwell's study, adheres to that convention. Blackwell also used another unit, the foot-Lambert where

$1 fL = 3.43 cd/m^2$.

The sky brightness values in the observation direction become when converted

$$L_0 = (1.5E-3)(0.1)(10000)600000 cd/m^2/892 = 1009 cd/m^2 \quad day$$

and

$$L_0 = (4E-10)(0.1)(10000)600000 cd/m^2/892 = 0.00027 cd/m^2 \quad night.$$

For the day-time sky brightness the threshold contrast varies over the range of interest down to a 0.73 arc min UFO as

$$Log C_T = -\frac{1.464}{3.005}\theta + 0.0129.$$

The UFO is resolved by the eyes over this range. Below 0.73 arc min the UFO becomes an unresolved point source where the threshold contrast increases as

$$C_T = \frac{0.187 min^2}{\theta^2}.$$

In the night-time the contrast is negative and the absolute magnitude is used. For the night-time sky brightness the threshold contrast varies with UFO size in arc min as

$$Log C_T = -\frac{0.746}{37}\theta + 0.573$$

where the UFO remains resolved by the eyes over the range of interest.

With the preliminary parameters determined the calculations can proceed. The observation point contrasts at the photopic and scotopic peaks are calculated as a function of vision range, i.e., 0.9km to 10km. The photopic result is equated to the threshold contrast for day-time sky brightness. The scotopic result is equated to the threshold contrast for night-time sky brightness. The θ angles can then be estimated. For a range and θ, the required radius of the UFO can be found. The results from this method are shown in the first Table.

R(km) and θ(arcmin)=2r/R with C	Day-time specular + diffuse	Day-time diffuse	Night-time specular + diffuse	Night-time diffuse
0.9	0.00067 (c=4.13E5)	2.6 (c=0.053)	54 (c=-0.307)	29.6 (c=-0.948)
5	0.00063 (c=4.67E5)	1.6 (c=0.174)	47 (c=-0.419)	29.4 (c=-0.956)
10	0.00061 (c=4.95E5)	1.3 (c=0.233)	42 (c=-0.531)	29.2 (c=-0.965)
VISION	(unresolved, + contrast)	(resolved, + contrast)	(resolved, - contrast)	(resolved, - contrast)

The day-time cases show how sensitive the UFO radius is to the size of the specular cone. The radius must drastically increase as the specular component goes to zero in the diffuse only case. At night, the radius decreases as the specular component goes to zero. Day and night, the contrast magnitude decreases as the radius increases. A UFO below 28 arc min at C=-1 cannot be visually detected for the night-time background. This corresponds to radii below 3.7m at 0.9km, 20m at 5km and 41m at 10km. As expected, the UFO radius must be large to offset the low night-time contrast magnitude. At 10km the largest radius is 61m (42 arc min).

9 RADAR

The RADAR is typical of those for aircraft surveillance and operates at 2Ghz at the boundary of the traditional L and S microwave spectra. Its wavelength is ~280,000 times greater than the vision wavelengths. RADAR being an active system is not sensitive to changing day/night scenarios. The RADAR antenna will search at an elevation of 30 degrees as it scans about a vertical axis at the observation point and detects in the geometry defined plane.

A detailed system engineering design example from Skolnik (chapter 13) is used as a base-line. Some relevant parameters of this RADAR in no particular order are: single hit probability of detection of 90%; antenna rotation rate about vertical of 5rpm (1 per 12s); false alarm time of 120s; pulse repetition rate 250/s; number of target hits per scan ~17; number of target hits processed per scan ~12; pulse modulated sinusoid waveform of $\tau=6$ microsecond width; receiver bandwidth ~127Khz; single hit signal-to-noise ratio ~25; processed signal-to-noise ratio ~2; receiver Noise Figure 1.5dB ($\sim 2^{1/2}$); antenna gain at 30 degree elevation of 18.83dB (76.43); Losses 10db (10) ; Peak power ~7.7MW; Average power ~11.5KW; antenna dimensions (1:2.5): ~1.7m high x ~4.3m wide; unambiguous range ~600km=c0.004s/2; minimum range ~0.9km=cτ/2. The unambiguous range is one-half the distance between pulses. The minimum range is one-half the pulse distance (i.e., leading-edge of the pulse returns just as the trailing-edge transmits). The minimum range is greater than the far-field beam formation range.

The atmospheric attenuation of 2Ghz radiation due to Oxygen absorption is $\alpha \approx 0.0013$/km at sea level (Skolnik). Water vapor absorption and refractive bending are not considered. The molecular weight, 28.96g/mole, of the Earth's atmosphere remains constant up to an altitude of 87km (Jursa). The attenuation to a significant vertical altitude h can then be approximated as $e^{-\alpha h/2}$. The attenuation to the same altitude, but now at a 30 degree elevation (60 degree zenith) is, due to secant scaling (Smith), just the square of this or $e^{-\alpha h}$. However h is one-half of R, so the 2-way 2Ghz RADAR attenuation to the UFO and back is $e^{-\alpha R}$ which is >0.8.

The RADAR range equation from the example can be reduced to

$$Re^{\alpha R/4} = \frac{182 km}{m^{1/2}} \sigma^{1/2}$$

where σ is the UFO RADAR cross-section in square meters and R is in kilometers. The cross-section is a function of RADAR wavelength and UFO radius.

When the UFO radius is much less than the RADAR wavelength the cross-section is determined by Rayleigh scattering (Rogers, Skolnik) or

$$\sigma = 4|K|^2 \left(\frac{2\pi r}{\lambda}\right)^4 \pi r^2 \qquad |K|^2 \cong 1.$$

When the UFO radius exceeds about three-halves of the RADAR wavelength, the scattering is in the Optical region (Skolnik) and the cross-section is just

$$\sigma = \pi r^2.$$

In between the Rayleigh and Optical region is the Mie region (Skolnik).

The RADAR range can now be iteratively found using the UFO radius estimate from the vision results. The results displayed in the second vision matrix Table again show the manifestation of a small specular cone.

RADAR and HUMAN VISION: Comparison Hypothesis for UFOLOGY

RADAR in vision detection matrix (km)	Day-time specular + diffuse			
0.9	X Rayleigh cs			
5	X Rayleigh cs			
10	X Rayleigh cs			
RADAR	<u>NO</u> RADAR DETECTION, small r<1mm due to small specular cone			

The small radii (<1mm) yield a small Rayleigh cross-section that does not allow RADAR detection at the vision ranges for the day-time specular and diffuse case. As the specular cone somewhat increases, radii below ~1cm should remain only detectable by vision. Mie RADAR scattering will probably dominate for 2cm<r<25cm. As the specular cone increases (more diffuse, less specular) UFO radii above ~3cm should become detectable by RADAR at or beyond all the vision ranges.

In the subsequent Table, at the remaining day-time and two night-time cases, the RADAR will easily detect at and exceed the vision ranges with the cross-section now in the Optical scattering region.

RADAR in vision detection matrix (km)		Day-time diffuse	Night-time specular + diffuse	Night-time diffuse
0.9		✓ Optical cs R=106km	✓ Optical cs 600km	✓ Optical cs 600km
5		✓ Optical cs R=354km	✓ Optical cs 600km	✓ Optical cs 600km
10		✓ Optical cs R=577km	✓ Optical cs 600km	✓ Optical cs 600km
RADAR		RADAR DETECTION >> vision range, 1.9m>r>34cm	RADAR DETECTION >> vision range, 61m>r>7m	RADAR DETECTION>> vision range, 42m>r>3.8m

The three lowest RADAR ranges for these cases are shown as 106km, 354km and 577km. The maximum is the unambiguous range.

10 THE UFO THERMAL EMITTED RADIATION

Now consider emitted light in addition to the assumed reflected light. The gray-body emitted spectral radiance from the UFO can be found using the Planck function (Zissis) to be

$$L_{\lambda,T} = \epsilon_H \frac{11911}{\lambda^5(e^{14388/\lambda T} - 1)} \quad W/cm^2/\mu m/sr$$

with λ either vision spectral peak wavelength in microns and T the temperature in degrees Kelvin.

The emitted spectral irradiance at the observation point is

$$L_{\lambda,T} \frac{\pi r^2}{R^2} e^{-(0.043/km)R}.$$

From the radiometry section, the reflected spectral irradiance at the observation point is found to be

$$\left(\left(\frac{2^{1/2}}{4} + \frac{1}{4}\right)E_{\lambda,sun} + \frac{3^{1/2}\pi}{4}L_{\lambda,0}\right)\rho_H \frac{r^2}{R^2}e^{-(0.043/km)R} + \left(E_{\lambda,sun} + L_{\lambda,0}\pi(\sin\phi)^2\right)\rho_s e^{-(0.043/km)d}.$$

Equating the emitted and reflected spectral irradiances gives

$$\pi L_{\lambda,T} = \left(\left(\frac{2^{1/2}}{4} + \frac{1}{4}\right)E_{\lambda,sun} + \frac{3^{1/2}\pi}{4}L_{\lambda,0}\right)\rho_H + \left(E_{\lambda,sun} + L_{\lambda,0}\pi(\sin\phi)^2\right)\rho_s \frac{R^2}{r^2}$$

and solving for T yields

$$T = \frac{14388/\lambda}{\ln\left[1 + \dfrac{11911\epsilon_H/\lambda^5}{\left(\left(\dfrac{2^{1/2}}{4} + \dfrac{1}{4}\right)E_{\lambda,sun} + \dfrac{3^{1/2}\pi}{4}L_{\lambda,0}\right)\rho_H + \left(E_{\lambda,sun} + L_{\lambda,0}\pi(sin\phi)^2\right)\rho_s\dfrac{R^2}{r^2}}\right]}.$$

If the UFO is in equilibrium with the troposphere, its temperature will be near 285K(53F) at 0.9km, 271K(28F) at 5km and 255K(-1F) at 10km (Jursa). If the UFO temperature is much less than T, reflection will be the dominant radiative transfer mechanism. Conversely, if it is much greater than T, emission will be the dominant radiative transfer mechanism.

The lowest T will occur for the night-time diffuse reflectance only case where

$$T = \frac{14388/\lambda}{\ln\left[1 + \dfrac{11911\epsilon_H/\lambda^5}{\dfrac{3^{1/2}\pi}{4}\rho_H L_{\lambda,0}}\right]}$$

or T=1020K(1376F). Thus, reflection is the dominant radiative transfer mechanism for all cases, as has been properly assumed.

The melting points of pure Al and Cu are 934K(1222F) and 1357K(1983F) respectively (CRC). The UFO would have to be molten, even approaching boiling, to have dominant emission detectable by human vision.

11 SUMMARY

The reader should become comfortable with these implementations using examples of their own and the results herein should not be over-generalized due to the large input parameterization. However, a class of small UFO radii (~3 centimeters or smaller) could produce an unresolvable specular reflection, depending on specular cone size, that is detectable by the human eye in day-time, but is not detectable by RADAR at all of the vision ranges. These radii would be "Under the RADAR" so to speak. Conversely, larger UFOs in night-time negative contrast that do not exceed a minimum sub-tense and radius will not be visible, yet will be readily detected by RADAR. Also, visual detection scenarios of appreciable emitted radiation are unlikely but not impossible.

It should be noted that a large night-time negative contrast detectable UFO may occlude some of the ~2000 unaided visible background stars seen from any point on Earth. This occlusion will probably result in a spatial non-uniformity eye/brain discriminant that enhances the probability of detection at the observation point. This was not considered when the spectral radiance of the night sky was assumed Lambertian scattered starlight. A similar effect should not occur for the day-time conditions.

A few comments on assumptions follow. Non-spherical shapes could be used, although coordinate systems and analytic geometry would need more constraint due to projected area and reflectance dependence on surface normal. Also, closed-form solutions for RADAR cross-section exist for other shapes. Atmospheric conditions could be changed and degraded due to weather, although RADAR would quickly dominate in the comparison. Other phenomenology could be considered, like highly ionized gases, such as the atmosphere in a lightning strike. Some RADAR sets can detect the ionization channel and human vision can detect the transient discharge. Again the point is, none of these are UFOs, but they could be used comparatively in UFO case studies.

Finally, with predicted results, the astute reader will realize the process is not complete without attempts at testing and falsification. Un-falsified hypotheses outnumber falsified ones. The falsified ones are abandoned or modified and recycled again as un-falsified. That's how science advances. In the end, science and perhaps even UFOs, always bow to reality.

12 WORKS CITED

Blackwell, H.R., "Contrast Thresholds of the Human Eye," Journal of the Optical Society America, Vol. 36, No. 11, 624-643, 1946.

CRC, Chemical Rubber Company Handbook of Chemistry and Physics, CRC Press, Boca Raton, FL.

Dudzik, M.C., Editor, Infrared and Electro-Optical Systems Handbook, Vol. 4, Electro-Optical Systems Design, Analysis, and Testing, ERIM/SPIE, Ann Arbor MI/Bellingham WA, 1993.

Fox, C.S., Editor, Infrared and Electro-Optical Systems Handbook, Vol. 6, Active Electro-Optical Systems, ERIM/SPIE, Ann Arbor MI/Bellingham WA, 1993.

Jursa, A.S., Editor, Handbook of Geophysics and the Space Environment, Air Force Geophysics Laboratory, USAF/DOD, 1985.

Kahneman, D., Thinking, Fast and Slow, Farrar, Straus and Giroux, NY, 2011.

Kean, L., UFOs, Three Rivers Press, NY, 2010.

Pigliucci, M. and Boudry, M., Editors, Philosophy of Pseudoscience, University of Chicago Press, Chicago IL, 2013.

RCA, Radio Corporation of America Electro-Optics Handbook, Harrison, NJ, 1974.

Rogers, R.R., A Short Course in Cloud Physics, Pergamon Press, NY, 1979.

Skolnik, M.I., Introduction to RADAR Systems, McGraw-Hill, NY, 1962.

Smith, F.G., Editor, Infrared and Electro-Optical Systems Handbook, Vol. 2, Atmospheric Propagation of Radiation, ERIM/SPIE, Ann Arbor MI/Bellingham WA, 1993.

Wilson, E.O., Consilience, Vintage Books, NY, 1999.

Wolfe, W.L., Editor, Handbook of Military Infrared Technology, Naval Research Laboratory, USN/DOD, 1965.

Zissis, G.J., Editor, Infrared and Electro-Optical Systems Handbook, Vol. 1, Sources of Radiation, ERIM/SPIE, Ann Arbor MI/Bellingham WA, 1993.

13 INDEX

Absorption 7, 13, Aerial 1, Aerosol 7, AIAA 1, Aircraft 13, Al 3, 18, Antenna 13, Atmosphere 7, 8, 13, 19, Attenuation 7, 13, 14, Azimuth 5

B-2 3, Background 10, 12, 19, Bandwidth 13, Beam 13, Bell 2, Black 3, Blackwell 2, 9, 10, Blue 10, Boiling 18, Brain 2, 19, Brightness 10, 11

Candela 10, Carbon Dioxide 7, Clouds 5, Coatings 3, Color 10, Columbus 1, Condon 1, Cone 4, 10, 12, 14, 15, 19, Conspiracy 1, Contrast 9, 10, 11, 12, 19, CRC 18, Cross-Section 3, 14, 15, 19, Cu 3, 18

d 4, 5, Detection 1, 2, 10, 13, 15, 19, Diffuse 4, 8, 12, 15, 18, Drake 1, Dudzik 4, 8

Earth 4, 5, 13, 19, Elevation 5, 6, 13, 14, Ellipse 8, Emission 18, Emittance 4, Encounters 1, Engineering 2, 13, Equilibrium 18, ET 1, 2, Eye 2, 11, 19

F-117 3, Falsification 20, Far-field 13, Florida 1, Fox 3, 4, FWHM 10

Gases 19, Geometry 5, Geophysical 2, Greek 1, Green 10

h 6, 13, 14, Hemispherical 4, Hewish 2, Hits 13, Horizontal 4, 7, Human 1, 2, 3, 5, 10, 18, 19, Hypotheses 20

Illuminate 8, Incident 4, Intelligence 1, Ionization 19, Iota 2, Irradiance 6, 8, 17

Jursa 4, 13, 18

Kahneman 2, 10, Kardashev 1, Kean 1, Kelvin 17, Kirchoff 4

L 13, Lambertian 6, 19, Life 1, Lightning 19, Lobe 4, Logos 1

Melting 18, Metallic 3, Mie 14, 15, Military 3, 5, Molecular 7, 13, Molten 18, Moon 5

Nobelist 2, Noise 13, Non-uniformity 19

Observation 4, 5, 8, 9, 10, 11, 13, 17, 19, Opaque 4, Optical 7, 14, 15, Oxygen 13, Ozone 7

Paints 3, 5, Paradigmatic 2, Phase 5, 18, Phenomenology 1, 3, Philosophers 1, Photometric 8, 10, Photopic 10, 11, Photosphere 4, Physical 2, 3, Physics 2, Pigliucci 1, Planck 17, Polarization 6, Polished 5, Power 1, 13, Probability 1, 10, 13, 19, Pseudoscience 1, Psychophysics 2, Pulsar 2, Pulse 13

R 4, RADAR 1, 2, 3, 5, 9, 13, 14, 15, 16, 19, Radiance 6, 8, 9, 17, 19, Radiation 6, 13, 17, 19, Radiative 18, Radiometric 8, 10, Rayleigh 7, 14, 15, RCA 6, 7, 8, 10, Receiver 13, Reflectance 4, 9, 10, 19, Reflection 4, 6, 10, 19, Refractive 13, Resolved 11, 12, Rogers 14

S 13, Scan 13, Scattering 6, 7, 14, 15, Scotopic 10, 11, Sea-level 7, Secant 14, Semicircle 8, SETI 1, 2, Signal-to-noise 13, Sinusoid 13, Skolnik 13, 14, Sky-shine 6, Smith 14, Solar 6, 8, Spectral 6, 7, 8, 9, 17, 19, Specular 4, 9, 12, 14, 15, 19, Sphere 3, 4, 6, 7, 8, 13, 18, 19, Starlight 6, 19, Stars 19, Steradian 10, Sun 4, 5, 6, 8, 9, Surface 4, 19, System 13, 19

Temperature 17, 18, Terrestrial 4, Threshold 10, 11, Troposphere 6, 18, Turbulence 7

UFO 1, 2, 3, 4, 5, 6, 8, 9, 11, 12, 14, 15, 17, 18, 19, 20, Unambiguous 13, 16

Vapor 7, 13, Vertical 5, 13, Visibility 7, Visible 6, 8, 19, Vision 1, 3, 4, 5, 6, 7, 10, 11, 12, 13, 14, 15, 17, 18, 19, Visual 2, 12, 19

Water 7, 13, Waveform 13, Wavelength 10, 13, 14, 17, Weather 19, Weber-Fechner 10, Weight 13, Wilson 2, Wolfe 4

Yellow 10, Yukawa 2

z 5, Zenith 14, Zissis 6, 7, 17

THE AUTHOR

Born in the rural U.S. from Germanic descent, H. David Umphenour was, until his retirement in 2013, a senior civilian research and development weapons scientist with the U.S. Department of Defense in California. His nearly three decade career in sensors, precision munitions and missile guidance at a laboratory dedicated to A. A. Michelson by Michelson's student R. A. Millikan spanned the end of the Cold War through Counter-Terrorism and Missile Defense eras. He earned an A.A. in humanities from Fort Scott Community College and a B.S. in physics and applied mathematics and a M.S. in physics, both from Pittsburg State University. His Ph.D. in physics from Pacific Western University comprised resident scholarship and research with predominate nonresident scholarship and research at the University of Missouri and Pittsburg State University plus career related collaboration with the University of Arizona. During graduate school in the 1980s he met one of the last giants of 20th century science, J. A. Wheeler, student of Bohr, teacher of Feynman, et al. A center-right secular humanist in the Enlightenment, Scientific and Renaissance traditions, he is a member of the American Institute of Physics Sigma Pi Sigma, a member of the American Geophysical Union and a senior member of the American Institute of Aeronautics and Astronautics.

www.ingramcontent.com/pod-product-compliance
Lightning Source LLC
Chambersburg PA
CBHW031559210526
45464CB00003B/1354